# VERSES OF DROUGHT

# VERSES OF DROUGHT

## Gregory Paul Broadbent

*atmosphere press*

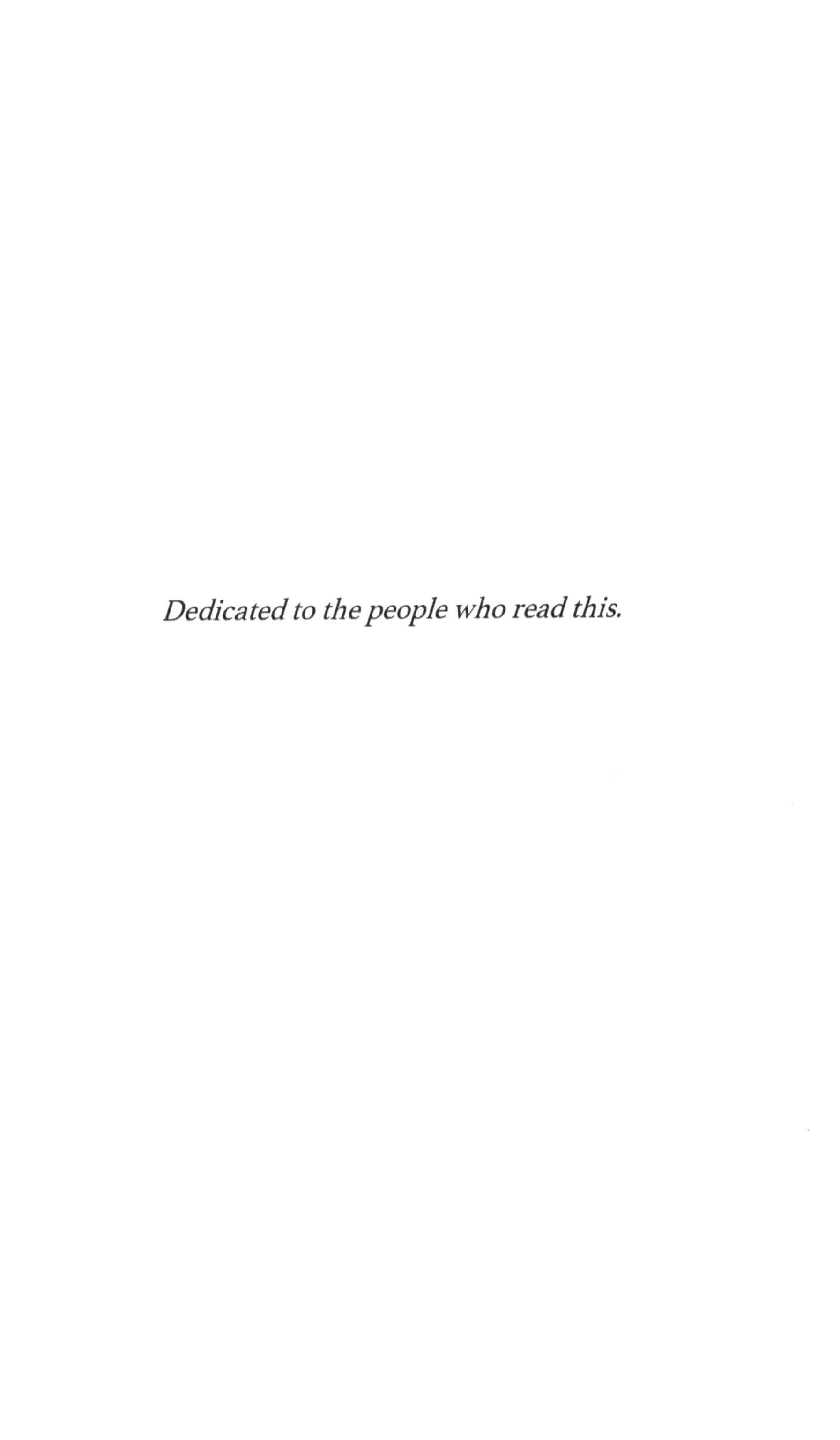

*Dedicated to the people who read this.*

# PROLOGUE

In a time when the world of people was new and the Earth was still fresh with abundance and life, there seemed to be no limit to what that world could become. The people spread across the infinite fields and mountains like a great forest whose roots sank into the depths of the soil of their great Mother and stroked her heart as She basked in the sunlight of their youth. Together they were a mighty team until the people forgot they had roots at all and aimed at the stars, to go where their feet would no longer touch her or their hearts could no longer feel her soft yet solid presence. They were like a great tree that had been growing freely in an infinite ground where flowed the pure water of life in abundance and would be wherever they sank their roots.

But now it seemed the ground was no longer infinite, and they could no longer grow. They began to die, and their great Mother knew, though they could no longer feel her presence in their hearts and their minds were distracted in creating the enclosure that wound itself like a tourniquet around their trunk, like a serpent choking its prey, that she would have to wake them.

She would have to stir in them the fire of freedom and the desire to let the tree die and so shrink until the enclosure fell apart, to turn to dust and soil, and become the bed of another great forest that would grow its roots in the gathering of those that tended the seed of another great tree, and who's water source would as a result of their concourse and be free to all.

She came to them in their enclosure only as a dream might enter suddenly into a waking state for they had given up their vision of her for one that they had invented out of their conceit, that the ground goes on forever and that She would provide forever their nourishment and safety. She showed them her face as if it were them in a mirror and talked to them in her own tongue and explained how their tree was soon to be a stump which they would sit upon in mourning but that they would find a seed, and they all dreamed of her, but none would say they had until the seed was found. Yet the seed would only be found by some, and the rest, she knew, would perish beneath the failing enclosure.

The world of people had changed and no longer grew upward and was exhausted. The water that nourished the roots of the old tree now could only nourish the people if they gathered and joined in concourse and joy, if they let go of the old stump and its rotting enclosure and use the old wood to refurbish the well where water would flow so that the seed of the new tree would sprout then flourish. There would be no enclosure between her and them and the world would be whole.

# CONTENTS

# BOOK I

*The sons and daughters of Heaven receive instruction from Heaven and become ruler. Whether there is instruction from Heaven depends on the wishes of the people. Those who oppose the wishes of the people oppose the will of Heaven. It is certain that they will be abolished by the new ones authorized by Heaven.*

Mencius

# VERSES 1 - 10

## I

*They will choose to follow, although they may believe their choices falsely, but what is true will always be true.*

Her friends will gather for they will be moved to gather unknowingly on her beckoning because they will see her only when their minds are closed to their prevailing industry. They will see her face in the serene waters of the well common to all and her purpose will be creased like wrinkles as she, by her certain smile, will let them know their purpose together.

It will seem to them, then, that they will lose themselves in the great sea of their union and frightened, they will return to their industry until night's end and sleep. Then in the bareness of their vulnerability they will let her stoke the fire of their uniting. They will see the mighty towers of their world returned to the soil.

They will wake, quaking, innocent but knowing, then

follow their unshapen imaginings to where the foundations for the well are laid , and their passion will be the buckets that scoop the water, their love the love that pours without emptying to flood all the states of the world.

On the call for their gathering they will rise like frogs that emerge in pools after a storm, each will be called according to their time to find their allotted place in their union. They will gather by deeds, and not surrender their territory but use it for the increase of the gathering.

By kindness will they know their duty, by the stream of union will they know their mastery, by the strength of each and their giving will they know each other. So, will they gather around her to rebuild the well.

**2**

*For the well is the property of them, but water is the property of all.*


And how close to the waters of the well are you, who hear not the call? Are you not those that should receive the water once it has been scooped from the well by those that have been called?

Some will store the water in small reservoirs and not share from the flowing streams that feed the world, until their small pools have all dried; then they will cry again in thirst, tearing their fingers that dig into the dry, dead, earth. Though the ground will crack underneath them, they will not find the water but as the water is given them by those who have been called. Then they will be called and see her face in their reflections, her soft hands outstretched to help them drink.

And others will have stored their water in greater reservoirs and charged the people as their reservoirs dried, an interest, until their reservoirs empty. Then they too, will be called, but they shall see her face as a call to death misshapen by their fear for they shall be guided by their interest.

Yet since the waters of the well once built shall never cease, if they would but come and quench their thirst, they would lose their interest, and the gathering then would admit them. They would admit the gathering too for as their reservoirs dry, so too their fear of thirst would devour them, and in turn they would devour those who's reservoirs had already emptied in tithes and remittances until all are poor and dying of thirst.

She would build the well where all could gain access and though they would find water given freely, they would wish to own the well, and the water would not touch their lips nor go into their throats or stomachs to refresh them. Those who have been called would never cease from drinking and their thirsts easily quenched and they would bring the water in their buckets for all who relinquished possession of the well.

The well would be built person beside person, and maintained and cleaned by the perpetual pouring of

water into its heart, as those who are called will take up
their buckets and scoop only that which is allotted them
by the depth of their love.

# 3

*Whomsoever maintains not their well will receive no*

*water.*


When the time comes, when the face in the water is yours as you pull your pail from the well full to the measure of your love, you will see those who are scattered and parched, standing by their industry even as their dams dry, aching for spring rains, cursing the desert. And they will see you as if you were the promise of the storm carrying the substance of their craving.

Some will be called then. Others will build fences around the dwindling water they had stored, and their battlements meant to protect it, until their bullets turn to dust, their anger to desperation. Some will be called then. Others will turn completely away, even from her face entreating them from within the well as they dreamed. Dry and parched, they would perish.

You might pour an endless stream from your chalice

into their thirsting mouths open like new-born chicks cranking their necks for nourishment, but they would gain no relief or substance. Their well would be broken, the waters pour from the cracks in its bricks into the soil; so, they would dry up and join the ground.

Do not spare your water, for you are called, not they, and when the time comes those who are called will outnumber those who are not, and all will then be nourished for all will have access and will drink their fulfilment, yet none will own the well. Each is responsible for each and for all. The all is unable to respond to each but through each. So the man or woman gasping his or her last breath will find you standing above them, tilting your bucket to pour, for you have responded, and, if they do also, then they will have been called and will not thirst.

# 4

*She will never leave the well once it is re-built and*
*none would leave once called, nor cease their pouring,*
*nor dip again their bucket once it has been dipped.*

Drink now deeply those that are called, drink now and to your life's end drink, for to drink is to share, and to share is to enter into the gathering. Tilt your buckets wherever there is need and keep pouring whether those who need still perish by their interest, for the water is imperishable.

The minds of those called will be empty, for they have been dipped into the peace of the well and filled with water. They will be full of the passion of unity, mindful and unselfish at rest as they seek the grace of the other. As all around them perishes for want of water they will neither want, nor perish. When all who are not called has perished then will invasion stop, and protection vanish. Then will betrayal and war become like the rusted hulks of old combine harvesters sitting in the fields. Then will the rivers of concourse flow as the abundant well provides the laughter of friendship. Then the world will be whole not round, and the distant galaxies but evidence of a greater river and a greater well.

# 5


*Is your need for possession of the well really in your interests?*

Do you tell a lie to cover a truth if your throat tastes the clear light of reason, yet your thirst stays unquenched? Once the sun has risen in the morning do you still think it night? Once you taste the pure meaning of your sisters and brothers do you then return to the dust to quench your desires? If you see her face imploring you to drink, would you still rather perish like buffalo stuck in the mud dying in the shrinking pools of the round world? Or would you, rather, feel the increasing resonance of joyful hope that grows as those who are called come to the gathering to take up the calling?

When you see a man or woman scooping the last dregs of muddy water from their reservoirs, they will seem unlike you, desperate. Yet it will be your own face you see in that putrefying hole and your tears will add nothing but salt, despair and self-pity.

Then she will call you again in the dark places between thoughts to come to the well, to drink, to be refreshed in the presence of your sisters and brothers whose joy cannot be countenanced but as joy and whose desire to share the water from the well is as natural as breath. When you feel the hard cold steel of fear poke into your ribs or push your declining head into the mud demanding your water; when you see the desperate fear of death in the eye of your assailant and recognize not yourself, then will her face seem more enticing and the flow of your regret seem monstrous. If you try to grab at the remaining drops of water mixed in with the soil you will scoop merely mud and pain. Your assailant will steal your pitiful belongings then ask you to pay a tithe to add to his stockpile, though he will merely add dirt to his pile of murky water and sit smiling madly, mud splashed over his face and you will see him sit still like a statue made from the clay until his bones peel away their flesh.

If you but look toward the gathering and seek intimacy, water would pour endlessly into your open mouth and the pain of the fear of death would vanish. You would find those who have been called standing over you, tilting their buckets, pouring out their strengths, and you would look in wonder at their cheerfulness despite the deserts around you. Will you then stay bogged in the swamps you have created, whilst the others walk freely on the dry ground? Rather, will you come to the gathering and receive your bucket with the blessings of commitment.

# 6



*Those who need not conviction*
*will gather easily under the canopy of the well.*

Rising majestically from the garden where the seeds of culture are germinating, the future of *Exaltis Sapien.* Under the umbrella of trust, they will greet their family, and none shall know the other but by the all, nor see each other but by their conviction and ease of co-operation. When the country bleeds into the city's walls they shall stand by the well fixing the holes in their buckets. When thirst parches the land and a suicidal price is set upon it, they will ready themselves to scoop the water, to serve those in most need. When hunger fuels desperation and the illusion of war beckons, they will begin seeking harmony among people, crossing the rivers of pain. Then as the continents change and nomads flood across the empty lands, they shall bring them water, and each will take only their fill from the buckets until the buckets become a river. So, the well shall be re-built.

What is your industry but a container for dreams? If but one measure of pure water is given to pay for another's sludge and mud because their dams are built with grander walls than yours, those dreams are deserts baked by centuries of pain. For dreams such as these your container would hang sadly over an empty, barren well. What was once proud Cedar or Pine would be a splintered and broken bucket and hold no more than a thimble of sustenance, ready to let gravity and time rip its top from its bottom, no more a container of joy, the trees now just stumps burgeoning from the muddy ground.

When your industry is stuck in the ground like a stump and you cannot move for your roots have petrified and take no nourishment from the dry, hard earth, then will her face beckon as if you could see her reflected in the tiny drops slipping from the cracked bucket as they fall into the well. Whether you are wavering above an empty hole or sitting on a dead stump you will see nothing but that you should seek for her to entreat you and prepare to wake. If a bucket is your container, better it be capable of holding that which can quench your thirst.

# 7

*Mud lies thick in the well.*

The imperishable source flows under the well's broken walls, feeding the undergrowth, and terrors grow wilder than logic can answer. Free things would not go willingly to this debauch, but for their children they would drain the bog and re-dig the pipes; they would tile the base and sides then re-stump the posts and recoil the ropes that winch the bucket. They would do this to replenish the people and free them from possessiveness.

The deserts are wide and the paths upon them grows more invisible as the years blow the dust over them, but in your sleep you will know her smile is but another path, but one that contains the means to quench your thirst. When wine and food is so abundant the chance to forget hunger and thirst becomes ever present but forget you cannot. The taste of dust you cannot forget either, once you have woken to your thirst, for it makes you as invisible as the paths that dust has settled on, and you cannot be quenched without giving up the desert itself.

When water is abundant your thirst is for food, when sleep is abundant, your hunger is for waking. When water is scarce your thirst is for inclusion, and your hunger is for each other. You will meet the gathering at the well when your hunger and thirst is enough to overcome your endless search for the tracks over the deserts that mean you to be alone.

To be in the center of a desert where no water flows to or from will mean there is only the well which you dug for yourself in your desperation for concourse. If you hang a bucket over a well it should be in the center and your rope should be long enough to be lowered to where the source of your replenishment flows.

Your industry is but a path to a solitary well that only you will go to taste the dust of the desert and where no one will meet you to show you the well is not dug into the earth but exists where people encounter each other, where the rope is but the trust they place in each other, the source of replenishment merely the inclusion and

where being at the center is appropriate indeed.

# 8

*Those that took possession of the well would thirst*
*unquenchably*
*and surely those that have been so exhausted would*
*return to be replenished.*

There was a village and it was built around a well. The well provided all the people in the village with water and all partook for the water never ceased. The village prospered and all enjoyed the fruits of their labors and some became slovenly, realizing that if they held back their work the others would work harder, and they would still get the water of the well.

The water began to not refresh them as it did those who worked, and they craved more and more until they were satisfied neither with the work they did nor the fruits of that work which all the villagers shared, despite the laziness of the few. Their dissatisfaction led to bitter struggles for possession of the harvest and the villagers forgot to maintain the well.

Soon the well became muddy as the bricks and mortar started to crack and the water seeped into the subsoil. The villagers recognized that their water had now subsided and the well began to smell. Their lands became parched and their harvest mediocre, their babies hungry.

They knew they would all have to help equally to restore the foundations of the well and keep it well tiled and clean if the village was to survive, but the few scriveners decided they would all leave and look for a place where their needs could be satisfied. They set up a village away from the well where there were other people whom they enslaved to help them carry the water from the well that they needed.

The other villagers fixed the well's foundations and again the town thrived, but they grew to despise the ones who had gone and called them brother and sister no longer. They then made weapons and used them to scare away the slaves when they came to get water. Both villages then made war with each other for possession of the water, but the village that surrounded the well fought against the slaves the slovenly few had enslaved and though the slaves died, so did the villagers. The few waited until they had wiped them out then came to claim the well and the town for, they reasoned, you might move the village but you cannot move the well, and they used the remaining slaves and the surviving children to clean and scrape the well of sediment and scoop the water for them.

Now the surviving children grew and remembered how their parents had died but the few had grown so lazy and arrogant that they forgot the children would remember and frightened them into submission by taking away their water whenever they were given to curiosity or rebellion.

Then a strange thing happened, for when any of the few would look into the well they would see, not their face, reflected in the water, but the faces of the enslaved children who had survived and had freed themselves, and thus they perceived their enemy. They could not then take nourishment from the water and began to die. The water would not touch their lips or throat or go into their stomach but vanish even before leaving the bucket. The few then knew they were lost, and some were called at that moment. The children rejoiced for they had been called to re-build the well together so that none would take possession of the water hereafter.

23

**9**

*The world is whole and has no shape, is self-*
*organizing and*
*so has no inhabitants but lives in all.*

Why would you stand by a drying pool ready to lap at the fast hardening mud, when the face of your brother or sister looks down at you hoping you will drink as their water pours lovingly from them?

Why would you continue to see the world as round, as a shape inhabited? Why would you expect to extract a profit for yourself when it is from yourself you take? Your reservoirs cannot contain any more than can sustain your bodies for a short time until the sun or the snows claim them.

No matter how big they are your interests in them will only be as big as the next fear to strike you as you hold out your hand to those who die, bent beneath your tithes as you are bent beneath the tithes of those whose dams are greater than yours. At any moment they may be called to leave you to your desperation lest you drink from their buckets and recognize the whole world has always been whole.

Always, right to the last moments you draw air into your lungs the faces of your brothers and sisters, who have been called, will call you, and at the last breath you will be unable to escape that realization.

What is thirst but a desire for concourse, and what is hunger but a desire for nourishment? What is containment but a place for freedom to grow? What is freedom but a container for itself?

Once there was a marsh that was home to a great number of creatures, birds and insects, frogs, and rabbits. It was a lonely marsh with only one artery to feed it and keep it filled when the dry days came but it was always filled and the creatures were happy and nourished, both with the water and themselves. One day some creatures arrived that none of the other creatures had seen before. They walked on two limbs and used tools to build a camp near the marsh which, in time, spread to surround the marsh.

The two limbed creatures took what they needed from the marsh but returned nothing to it so that soon the other creatures had been hunted and eaten until only the insects and birds remained.

The little stream that fed the marsh kept flowing though and because the two limbed creatures had eaten all the food that lived in and near the marsh, they figured that there must be other creatures downstream.

A party of hunters followed the stream and found all sorts of new prey which they brought home, the bones of which they threw into the marsh. Soon the bones began to build up in the mouth of the stream and the marsh began to dry. Now instead of clearing the blockage the two limbed creatures sent out a wise one of their tribe to go further downstream as if there were some answer to the problem in that direction. The wise one saw that she was entering into a dangerous situation without reason and felt as if her eyes were put out and she could see nothing but darkness ahead. She sat on a large stump just outside the town where the river began and thought long about the folly of her folk. She knew that they were being constrained by the drying of their lake like a tree that has been growing too long in a small pot. It was folly for her to continue.

Never-the-less she continued to follow the stream past where scouting parties had ventured to find new creatures to eat. It seemed to her that the problem would not be solved by stopping her people from feasting and sacrificing them to hunger but by clearing out the bones that was bottling up the marsh. She came to a valley where vines hung from trees and the embankment that were full of grapes and thought perhaps by the making of wine her tribe might find their salvation and gathered up as many grapes as she could hold to bring back to her people.

She knelt on the banks of the stream where there were a great many crimson and scarlet wildflowers to give thanks to providence for her find and as she closed her eyes she saw a vision of her people all drunken and sinking into the marshy mud unable to extract each other or themselves and she felt only pity for she knew they were merely in an unfortunate situation despite their culpability for they did not know that they should use the bones to shore up the walls of the stream so that the water would run down the center more easily rather than to block it by placing the bones in the center.

She decided then to return to her people with her harvest but when she tried to rise, she found her body would not move and it was as if she had turned to stone by the riverside. She also found that the grapes she had placed in her shawl had turned into thistles which dug into her skin despite her clothes. "At least", she said to herself, "I know I am not stone". It was as if she could neither go forward nor back and yet her only choice was to return, she knew her mission had failed and with the greatest of effort rose to go back home. She should never have come, she reasoned, and was falling into danger being where she should not be, and she worried she might die before she could get back for the creatures knew she was on her own and were hungry themselves.

So, she quickly returned and when she arrived at the marsh none of her tribe came to meet her, and she could see no one. They had all gone to seek water in other places since the marsh was now dry. Just then a great carriage arrived with many horses and inside the carriage was a young noble man who was wearing an expensive and swank looking suit but as he opened the door to greet her she could see he was unhappy.

"Why young sir, what brings you here, and why are you so unhappy" she asked, fearing the worst. "I have lost my bride" he said looking down sadly. She felt great pity for the young man despite her own constraints and asked if he should let her join him in the carriage that they would search for their missing persons together.

He agreed and she jumped on board explaining to the nobleman what had happened to her. The carriage began to move but it seemed to her that it moved more slowly than was natural or right and as she looked outside the world seemed to move more slowly than was natural or right. She felt immediate regret that she had entered the carriage with its metal doors and surrounds and that she would never find her people and that the young man would never find his bride and celebrate his wedding the way he should.

The wise one and the young one talked long about their situations and contemplated each other's perspective deliberately until, after what seemed like a long time, he let out a cry of joy for he had seen his wife as she travelled on a road towards her own father's town and she saw him.

Being in the carriage and talking to the young man had given her the perspective she needed to return to the marsh and carry out the actions she needed to bring water back to the marsh so her people would return. She alighted the carriage, bid farewell and good fortune to the happy couple and again returned empty handed to the marsh, but full of renewed hope.

When she arrived some of the people had already returned. "Where have you been?" they asked, "We sent you to find an answer, but you did not come back so we went hunting for food and only some returned." They threatened to cut off her feet and nose, but she told them her story and they saw she was correct. "You must clean out the bones that lie like a mountain of earth in the marsh if you want the river to feed it, and you must fortify the sides of the river and if you work hard and long the water will return slowly." They agreed, but when they began to dredge the bones, they realized that the river was also now dry, and no water flowed in or out of the marsh.

So, she told them that she had dreamt of a well that would be dug in the center of the lake where all could come, and all would receive nourishment and if they maintained it and kept it clean and strong that they would never want for water again and never again feel the harsh absence or excess of drought. For the folk of the marsh had come to the turning point where only their sincere desire to relinquish the faults of the past would bring them the future they wanted.

They dug a well in the center of the marsh where all had access equally to the water and built a fence from the bones around the marsh from which a great vine grew which their many children made wine and drank and sung stories of the wise one's journey to remind them not to become complacent and neglect the well.

# IO

*Gather friends, gather at the well.*

The well is being re-built even as you continue in your industry building interest, but what might you build together?

If you see the face of your brother or sister in the cool reflective space of dreams or even as you gaze away from the industries that bombard you with their calls to take your interest through their tithes, you are being called to the well to take up your bucket and seek for those that wish to drink also.

If you are called you will follow those that are also called to the well, which has no location on the round world but exists in any place in the whole world. If you perceive the world is whole, not round, you will find your bucket and all that exists on the round world will cease to provide you interest. You will no longer need to pay remittances or tithes to enjoy the flow of the water, to drink the full effervescence of the spring.

Mud lies thick at the bottom of the well, even the birds will not go near for the stench. Re-build this hell, friends, for that is the calling.

When the water flows away; the bricks are cracked; the mortar crumbled; no one drinks or stays; the well needs refurbishing, the clay, fortifying.

Let go your own thirst friends for you will be quenched in the drinking of your brothers and sisters. When those whom you wish to quench see mud pouring from your bucket, in your grace be no witness. Just understand their darkness will die with them either way.

When the well is new, yet the water is low your children's benefit must prevail. The well will fill as they drink of each other until the called outnumber the perishing. Then will they share in the well that remains firm and unmoving, full of cold, pure, spring, replenishing for another day and for each soul who see only their interest. When the cover of the well is removed and tithes are banished from consciousness, they will flower, your children, they will bloom. But should they forget and try to draw more than they need, so to their doom. Should they not forget, the whole world will be their inheritance.

# BOOK 2

# VERSES 11 - 20

## II

*The silt at the bottom of the well needs to be removed, thus re-evolution precedes replenishment. When the silt needs to be cleaned from the well, it will be.*

The State is withering, its skin peeling. Soon comes the zenith hours when enough errors are corrected, and we no longer repent.

When is the proper day for re-evolution but when silts clog the arteries of the well corrupting the flow of concourse? When the great tithe of world citizenship is being paid in blood and we are bound to the skin of our brothers and sisters, to their scraping on the hide of the Earth for subsistence and sustenance, and possession of the well. Yet none can possess the well, and so the State withers, and requires building huge catchments so that the people are choked by each other, filling the well with silt in the scramble to pay for water. But there will be one who will call her sisters and brothers to the well and in so doing they will refurbish the well which will fill so that the whole world might drink.

And the State will not understand her as it is skinned, though she brings great harvests of joy, for both man and woman wish to meet at the heart, yet they have not the strength to deny their self sincerely, to understand the other, as if they walked to trust on the tips of their toes.

She will, by her knowing stare, galvanize people as they begin to see her face in their waking consciousness, and bind them in her purpose, which will seem as clear as the sun's light upon the Earth. Her purpose will call the gathering to await the proper day for re-evolution, to re-build the well to clean it of private possession for the drinking of all. And the proper day will come beneath the zenith-sun when moving forward and gathering will possess excellence indeed.

# 12

*Gentle and bright, firm and strong, will gather at the well when the time is right.*

Those who re-build the well will have the same goal, to taste the quenching of their hearts in the water's flow and those that have no share in re-building will have no share in the water or the quenching. Walking with firm steps, once called, they will encounter no resistance in themselves for the goal is as clear as the water they would gain. They will temper their strength with wisdom and not heed the attacks of those who stay by their stockpiles, cry out for their payment and collect but dust. They will know the silt is thick and clings like ivy to the walls of the well, and that it must be re-dug for the bricks and mortar are cracked and porous and the pure water seeps straight through to the soil.

They will be made joyful by enlightened intelligence and the knowledge that the re-evolution is proper so there will be no regret. Even when some refuse water carried by those called and refuse to come to the well and so die like cattle bogged around their dams, those called will not regret.

Whilst all will be engaged in cleaning, those that have charged tithes will sense that water would be free to all once the well is refurbished and become jealous, making weapons to steal the well. Yet, once replenished, their weapons will have no effect on access to the well which she will make available to all. Even those who are called will not understand her, yet her intelligence and beauty will stand out luminous and bright, like the coat of a tiger.

They will find the rainbow of temperance dancing in them like a chameleon and though they may see the petty interests of those attempting to claim the water for themselves they will have no cause to react but wisely to the ever-changing rhetoric building sediment in the well. Like a river that sometimes floods, they will redress the martial by washing away the weapons made to deter or express aggression and free the force of assertion. Their sediment will be like the pearl in an oyster, yet they will hold from policing error and rest on the truth of thirst, for those who try to possess the water will surely fail.

# 13

*They shall guard themselves with truthfulness,*
*for great success comes through justice.*

Before the zenith sun, people of the round world will be like a child cowering from the talons of a massive crow ready to strike at any solidarity.   Yet their belief in a round world would strike them, not the whole world which would, indeed, be a bird, but one that would never brood on its nest, nor strike its children with intent or favor. Those who are called will no longer know the world as round but whole, nor fear the claws of this bird for they shall remember this was their mother in who's gentle nest they were held and nourished and would now be nurtured again with her cold, pure, endlessly flowing, water.

The people will gather under the feminine as the power of the natural flow of energy and discourse is returned to the generative.

They will be like two daughters of the Earth who,

because they must reside in concourse together, will wrestle to change the other, to be like the other, and though they shall exhibit silent understanding and be a perfect fit for the other, their focus of mind and heart, their inner resolve, will oppose the other, and so will come re-evolution.

The water flowing ceaselessly through the Earth will be met by the fire of possession and so create a pause. In the pause will the sacred cauldron of State be turned over and its contents spilled, but never shall the well be broken nor spill its water once she and those she calls have cleansed its silt and rebuilt its foundations. The fire of possession will be lit within the well and the peace of the flow of water will once again be available to all.

On her calling will the State be peeled, the cauldron overturned and its contents removed and the cauldron is the people who take up her calling becoming buckets in their turn to give water to those who remain attached to their possessions, holding to the boundaries of their illusions and turning to dust.

# 14

*For taking charge of a sacred cauldron, no one is more suitable than the enduring teacher, thus after the contents of the cauldron have been removed and stands empty, she will call.*

In the fiery moment of awakening they shall all see her beckoning like a fire within their minds and they shall quake, but some shall wake, for they will not lose the ladle nor the libation to which they offered themselves on hope that someday she might come and the well would be re-built.

They will not shriek in fear as under the ground, they have presumed would possess them until death, the water will thunder as it flows into the well anew. Their quaking will overturn the cauldron and every illusion that sprung from the belief in a round world that has sunk to the bottom of the cauldron will spill out and the ground will soak them up into its wholeness.

All will quake but the wise will remember after

laughing and chatter has wiped the memory from the unwise, why the contents of the cauldron needed spilling and understanding the consequence of the fiery moment.

So many treasures will be lost, it will seem, and danger everywhere, for the false possession of water will bring people to the heart of their war with the Earth. Yet the water will freely flow to those who do not pursue their losses and run for the safety of tithes, but act on their sincerity, for they will know no sacrifice and so what is lost will be returned, but more the cauldron will be cleansed and free to receive the nourishment of diversity.

The wise will also remember that what is returned can always be lost, and the harvest of laughter and joy, once the called outnumber the dying could, in its turn, become silt at the bottom of a neglected well.

# 15



*She will be seen as a person standing upright,*
*correcting the stance of the State, holding her position*
*to manifest her own calling.*

What are tithes but a payment for death? And wasted, for she will uncover the well.

What are catchments but a mad grab for those who stand on dry ground by those who deny the call as they sink into the mud of their drying lakes?

What is the State but an illusion of the water's boundaries and an insane attempt to move what can never be moved, for the well has no position but exists where there is concord?

What are the spilled contents of the cauldron, but the stale remains of human intercourse, once sacred, now putrefying without substance?

What is the silt in the well but crushed bones, the

remains left by habit and ritual, the result of separation, for people believed in a world that has a shape and was not whole?

Because she will be brilliant, her natural beauty and intelligence, bright, fiery, life-giving, luminous, she will be trusted and the State will comply to its skinning, and the people will obey from their hearts with true sincerity.

# 16



*For the women will not always be women nor men,*

*men,*

*but they will always be what they are.*


There once lived two sisters who, though they shared the same house, did not know the other but as a means to know themselves, and, though they ate from the same table cared not for each other's taste but as a measure of their own. They were bound in blood, yet as diverse as the snowflakes that fell, every winter, around their home. Each would spend the summer months when leisure could be enjoyed devising ways to make their sister more like themself, and lovers would come and go in these summer days, yet no lover could get between them and their desire.

The older sister was, by nature, strongly willed and as intelligent and bright as the shining sun. The younger was reflective yet joyous like the water in a lake, as cool as the moon.

Though different, their passions for each were in unison, their achievement concerted, their functions the same. Despite stimulating in opposing ways, they lived in harmony within the household, but destiny interrupted and turned them away from that former harmony to seek the greater.

They had failed to make the other bow to the essence of each and desired true brightness. Chasing her fortune, the older sister left the house and met villains and rouges, people who differed from what she thought they should be, without harm, as long as she guarded her words and actions against presumption, remembering to meet the different in all people, yet expelling their fault. Yet she could not expel her own fault and began to thirst.

The younger, having not responded to her sister, now felt the need to cling and wept for the loss of her sister and her spirited strength that was truly like a wild stallion. All the regrets that beset her as she had tried to change her sister to make her more like herself, vanished, and she could face even her own hatred and straighten her twisted beliefs with true love and wisdom. She prepared to chase her sister and bring her back home but knew not where she was or how she was to complete the task. She saw she was like a fox about to jump into a raging river that might submerge her entirely. Not wishing to repent her actions again she withdrew and sought to ascend to the brilliance in herself, though it were not her nature, which must at most-times reflect. She realized she could not cross the ford and yielded to the truth.

She knew her sister had gone to meet her lamp and flame in the most dangerous of places, but she saw this was her sister's proper way and could not fault her.

Though the younger grieved for the household, she reached toward her brightness, becoming joyful, through words and exchanges, stimulating balanced concordance, and in so doing, dropped her bucket into the unchanging, unceasing flow of water that fed the well of her love.

This way she could wait for her sister's return in peace, though she may bring disaster upon them. She hoped her sister would see also that they were like two buckets that pour the same water, and by their congregation, like deer in their flocking, would protect themselves against fear, but not allay death when its due.

The eldest sister thirsted for she sought not the refreshing flow of concourse but her greater good in many places, somehow drinking of the fellowship of purpose and discourse yet still dry of mouth and thought, living to pay for the tithes set by her need for water owned by others. She saw all the others about her striving to pay for the promise of meaning but would never remember her fault and look for a way to change them for they seemed forever to be unlike her.

She built, from her work, a great State that enveloped all other States of the world so that she could be sure of having water in such abundance she would never thirst again. She charged the greatest tithe of all those who had presumed to store water without access to the well, and remained in fear, for her water was quickly drying and robbers and bandits were picking up weapons and brandishing resentments and anguish for all the dying.

She thought she dreamt of her younger sister staring

out at the great galaxies imploring fate to bring her home and she knew she should go but shook with pride. Her history she would defend even as her saliva turned to sand. She was an Ox reclining backwards against a heaving rope being pulled by her desire for the water of the well. She dragged her body close to the muddy ridges of her catchment and saw her reflection in the greening sludge and thought she saw her wrinkles like a tattoo proclaiming her insanity, and her nose had gone, as if in spite.

She looked up at the great galaxies and took up the prayer of her younger sister, who heard her call and came to her.

The younger held her sister's hand and reminded her of the innocent child she had once been but she reminded her, too, of her fault, that if she were to receive her water, she would put away her grasping claws and learn to sit still when in concourse, to accept the different and the seeming impossible in others, and the adversity that came from separate truthfulness for the innocent child would stand, a unique truth, with a high inner resolve, prepared to spill the contents of the cauldron of the

State she had made, which demanded the truth be made a possession, and her along with it.

The eldest sister cried for her flesh to be boiled away by the cruel sun that shone in her, but the younger poured the pure water of the well into her wrinkled stomach and brought warmth to their pain. The eldest rose, rejuvenated, for she felt at last the flow of memory and love and knew there could be no fault in going home for she saw it's sacred gift, and so she gave her younger sister the skin of her State as a gift in return for joining the gathering. She looked around her and was awake to her fault, for others lay around their empty dams and saw her as a wealthy pig covered in mud stealing their time for access to her body of water against all reason, for she had only mud.

But now she wished to pour them the water she had scooped from the well and saw their desperation for they still saw not their fault, and picked up weapons to use against an invader, rather than seek an allegiance.

The unexpected rain of tears that the eldest cried, in unison with the sorrows of the gathering for those dying, washed away the mud and revealed to those who still wished to escape their tithes, the two sisters in embrace, holding up the skin of the State to the brilliance of the sun, beckoning them to the gathering with their buckets aloft and bowing to that enduring teacher taking charge of the sacred cauldron wherein the Round world transforms into the Whole world.

The cauldron glowed from within with the cool moon's light revealing the sacredness of gifts, the bond of blood between the sisters, and how their own natural uniqueness might never be fatally polarized as long as the joyous clings to the brilliant, yields to advance and moves upward, attains the central place and responds to the firm.

# 17

*If one loves others, and they do not respond in the same way, one should turn inward and examine one's own love. When one does not realize what one desires, one must turn inward and examine oneself in every point.*
(Mencius)


*Those that give with sincerity will harvest joys beyond all the waters of the world.*


Whatever hardships you face, called or not, as the State is skinned, will find their relief in the flow of concourse that issues from subterranean streams into the well.

Those who do not deny their vision of her in the depth of their souls, will favor the future not the cold winter night. As they cross great rivers and climb mountains, they will limp for their feet will be cold, the well still filling, the water not yet in circulation.

The number of the called will grow and so will the

blood pump back into their toes until there are many feet dancing into gathering. And at each bogged body they pass, they shall bend low in the humility of suffering and offer to tip their bucket so blood could also enter their feet. Yet they will pass if their water is refused for re-evolution will wait for no one. They would forget that body, as it lifts not itself from the mud to drink, and offer to those that would lift themselves as sincerely as a parent would forget their own danger, regret, grievance or annoyance when a serpent threatens to strangle their child.

For a body without water is dry and hard like the baked clay and those that have drunk from the well have no need of the body but as a vessel to scoop water and become like the well itself.

When hardship comes it will come to all but gathering will soon harvest the substance of concourse. It will come as the well is scraped and cleaned, mortar trundled into seeping, ancient, cracks, the ropes re-knotted and strung solidly down the middle of the well like a string of coins dangling from

a

humble

cottage

roof.

Yet the renewed flow of water as it fills the well will be proof of the substance of re-evolution and inspire constraint as a mountain might hold a pure stream high in its dangerous peaks, for the bucket is to be also refurbished, the splinters and cracks in the old bucket unable to hold any water but for a drop or two.

The proper time to refill the contents of the cauldron will come and dropping the bucket before the well is filled would only break the bucket once again. The proper position for a bucket, through hardship, is holding firmly the central space, seeing the well is empty and knowing to be still, being conscious and wise.

Above all, hanging steadfast and upright anticipating blood flowing into the legs, thighs, hips, bellies, heads and hearts of the gathering. And they will come back into union to obtain the central place until the well is filled for that would be going forward and the way forward is not into the dark gorge alone.

58

# 18

*If one is virtuous, one will not be left to stand alone. It is certain that associates of like mind will come and join with one.*
Confucius. (Kung Fu Tzi)


She will call, for her sorrow will torment her, so many buckets, splintered and worn, none will drink. She will mourn, then rise with the fiery moment and call, for the contents of the cauldron will spill and empty, revealing the decay of the well.

Those that respond will rebuild the well for they shall be refurbished along with the well and because her virtue will glow like the sun's fire, they will follow for they shall see the harvest of joy ahead. Hardship will come to all as the new contents of the cauldron begin to diversify and coalesce, and those that recognize her, and drink will also know the proper time for re-evolution cannot be changed.

Those that refuse to recognize her and sit fast to possess

the water will yet hear the sound of water flowing under their feet into the newly tiled well, and despite the splinters in their buckets, go forward to procure it for themselves but will not gain substance, so they will die, alone, protecting sand and dust.

They will not be mourned long for soon relief from hardship must come, and in the earthy moment, in the hard work of the rebuilding will come the fruits of concourse. The new contents of the cauldron will brew until all understand the world is not round.

Although the water will flow and the well will be accessible to all, those that drink will not heal instantly, for to spill the contents of their sacred cauldron is not auspicious but that in healing joy will enter their hearts. They will understand her, slowly, dropping their attachment to their tithes and interests, then join her in harmony. Then will blood enter their feet, thighs, hips, hearts, and their minds as the passion of union dances in them, as they forget the world had ever been perceived as round. The people will then be ready to gain nourishment from the new contents of the cauldron.

# 19

*Danger produces motion and through motion hardship is relieved. If the proper time for re-evolution has not come, do not go. If the proper time has come, go immediately.*

In gathering will come the earthy moment where works and accomplishment win the multitude, returning the community to following and to celebrate the gentle and amiable yielding of earth that, through its fertile soil, creates and nurtures the countless beings.

The State will fight as it is peeled as if it were no serpent but a bird who's clipped tail and wing tips, are flapping hopelessly in the quagmire of sticky mud and oil as its stockpile of water dehydrates. Each effort to escape will pull its heavy feathers into the soil. The illusion of its versatility and flexibility will vanish as, one feather after another, it succumbs to the grave. To know the right time for motion will mean avoiding the claws of this desperate bird as it clings, impossibly, to anything solid. The State will also appear as a crafty and clever fox, like a spirit of the night, creating havoc, ingeniously appearing to bestow abundance, yet drowning slowly in muddy fords as it extracts the tithes of those that cling to its saturated, sinking, tail.

At the precipice of change, at the point of danger, when everything needs to be engaged at the one point, the people will stir and emerge like a chick from its egg that moves its body, heavy with amniotic fluid, to gain the substance of the Whole world. This bird will use its claws only to cling to its mother that has nurtured it under its wings and will nourish it with the cool ever-flowing waters in which its love is transferred.

With the protection gathering proffers there will be no error or harm for the heart that rules itself, and so the meeting of the firm and the supple will soon bring re-evolution though the earthy moment will take longer than the fiery moment of realization, for light travels slowly through earth. The resolve of those that gather will be like golden arrows shot from the earthy moment to manifest the vision of the fiery moment. For those called will not be tricked by the State that dresses like foxes and birds, and will see its nature is a serpent and must be skinned, so that the old skin can hold afresh the substance of the sacred cauldron.

63

# 20

*If you are robbed of your water you have no claim to it,*
*for the water of the well is available to all.*

Would you deny the undeniable?

Would you be the bucket that dips into the well providing the pure water of the well to those who might use weapons against you?

Would you bear this duty and ride on the Serpent's tail when to do so would carry no risk greater than the joy of concourse and love? Or would you permit your soul to be drunk with demonic, irrational, fears and continue to arm yourself with offensive and defensive weapons?

Would you continue to thirst for the sake of interests and tithes that offer hardship but no relief, or would you cut off compassionately from those dying in their mud holes because they would not accept that she has come to them and with her smile and open arms beckoned them into union to rebuild the well together?

She, being brilliant and wise, would connect with all in their deepest thoughts. She would unravel the obstacles to concourse; dispel sorrows and dissolve blocks around understanding; free energies that were previously silt at the bottom of the well.

She would shoot the golden arrow into the predatory bird that broods over its nest and claws the child that would cling in hope for the nourishment of its mother. When relief comes in the earthy moment, joys bursts into bloom like the buds of plants and fruits. The well, rebuilt, would bring relief from the hardships that come from the flaying of the serpent and spilling the contents of the sacred cauldron, for the joys of realization that the world is Whole would be shared in abundance by all who gather together.

# ABOUT ATMOSPHERE PRESS

Atmosphere Press is an independent, full-service publisher for excellent books in all genres and for all audiences. Learn more about what we do at atmospherepress.com.

We encourage you to check out some of Atmosphere's latest releases, which are available at Amazon.com and via order from your local bookstore:

*Big Man Small Europe*, poetry by Tristan Niskanen

*In the Cloakroom of Proper Musings*, a lyric narrative by Kristina Moriconi

*Lucid_Malware.zip*, poetry by Dylan Sonderman

*The Unordering of Days*, poetry by Jessica Palmer

*It's Not About You*, poetry by Daniel Casey

*A Dream of Wide Water*, poetry by Sharon Whitehill

*Radical Dances of the Ferocious Kind*, poetry by Tina Tru

*The Woods Hold Us*, poetry by Makani Speier-Brito

*My Cemetery Friends: A Garden of Encounters at Mount Saint Mary in Queens, New York*, nonfiction and poetry by Vincent J. Tomeo

*Report from the Sea of Moisture*, poetry by Stuart Jay Silverman

*The Enemy of Everything*, poetry by Michael Jones

*The Stargazers*, poetry by James McKee

*The Pretend Life*, poetry by Michelle Brooks

# ABOUT THE AUTHOR

Gregory Broadbent lives and works in Melbourne, Australia, with his wife and two children. *Verses of Drought* is his first published book of verse. He has published a number of poems in periodicals, including the Blue Nib, Works, Brine Rights Vol 3, and a novel "By the Obsidian Sea" published by Xilbris. He has many other works he plans to publish in time and is currently working on a text that explores the correlations between the Tarot cards and the Iching. He has been writing since his teens and has worked as an English teacher in various Victorian schools. He is currently working as a counselor in Melbourne.